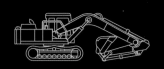

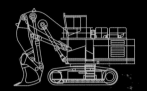

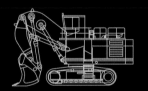

Book design by Annie Tsou.
Typeset in Eurostile and Helvetica Neue.
Manufactured in China.

Library of Congress Cataloging-in-Publication Data
Big, bigger, and biggest trucks and diggers.
p. cm.
ISBN 978-0-8118-6432-9
1. Caterpillar trucks. 2. Earthmoving machinery.
TL230.2.B54 2008
629.224—dc22
2007036530

10 9 8 7 6 5 4 3 2 1

Chronicle Books LLC
680 Second Street
San Francisco, California 94107

www.chroniclekids.com

BIG BIGGER AND BIGGEST

TRUCKS AND DIGGERS

Written by Erin Golden

chronicle books·san francisco

TABLE OF CONTENTS

BIG...BIGGER...
AND
BIGGEST
OF ALL!

Do you like trucks and diggers? Big ones? Powerful ones? Well, here you'll find all sorts—trucks that help build houses, fix roads, sweep snow, clear away messes . . . and more!

In this book, you'll explore trucks and diggers in order of size, from the smallest on up to a truck so mighty that it can carry the weight of 46 elephants! Check out the parade of trucks and diggers below—it shows the trucks and diggers you're about to meet, so you can see how the smallest compares to the most enormous.

Each of the machines you'll find in this book has special strengths.

SKID STEER LOADER   **MINI-EXCAVATOR**   **TELE-HANDLER**   **BACKHOE LOADER**   **MOTOR GRADER**   **ARTICULATE TRUCK**

• The **MINING TRUCK** is so big it can't be driven on regular highways, because it would damage them. It has to be shipped to a mine in parts on a train and then assembled on-site.

• This **WHEEL LOADER** is so big that its wheels alone are 12 feet (3.7 meters) tall!

• This **ARTICULATED TRUCK** is strong enough to carry around 31 tons (28 tonnes), which is about the weight of 52 polar bears!

• This **MINI-EXCAVATOR** sure looks small, but it's so strong, it could lift the weight of three cows!

• The **SKID STEER LOADER** can be a street sweeper, a snowblower, and a jackhammer, too!

• The **BULLDOZER** can use the teeth on its multi-shank ripper to bite into concrete and tear it apart!

To find out all about these vehicles and many more, just turn the page and *dig in*!

In case you were wondering . . .

The vehicles in this book are called "Caterpillar" machines because one of the Caterpillar company founders, Benjamin Holt, came up with the idea to replace wheels with tracks. When a photographer was asked to take pictures of these tracked machines crawling all over the ground, he remarked that they looked like gigantic caterpillars.

These are early Caterpillar tractors.

• The **SOIL COMPACTOR'S** tough steel wheels are perfect for packing dirt so that new pavement can be laid down.

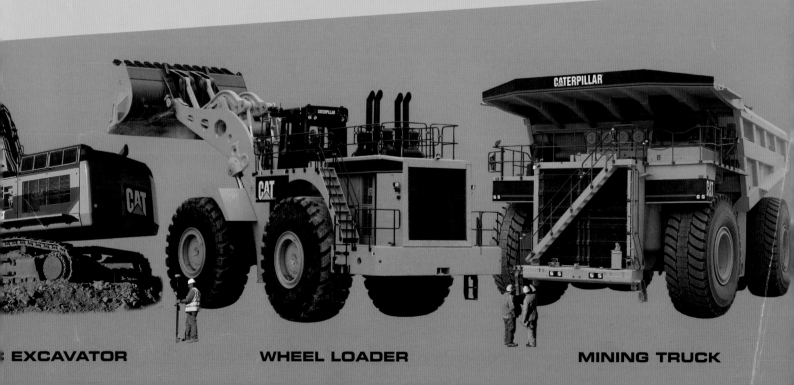

EXCAVATOR　　　　　WHEEL LOADER　　　　　MINING TRUCK

• The **BACKHOE LOADER** can do twice the work of most machines because it has attachments on both ends!

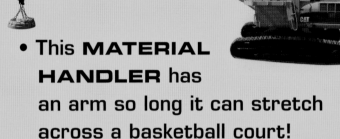

• This **MATERIAL HANDLER** has an arm so long it can stretch across a basketball court!

• The **MOTOR GRADER** makes the ground flat for a new road—but it can also be a snowplow!

• This **HYDRAULIC EXCAVATOR** can dig a hole 31 feet (9.4 meters) deep!

• This **TELEHANDLER** can lift a load to the top of a three-story building!

SOIL COMPACTOR   MATERIAL HANDLER   BULLDOZER   HYDRAULIC

The **SKID STEER LOADER** may be the smallest of the trucks in this book, but it's a very useful machine. This nearly 6,000-pound (2,722 kg) vehicle moves quickly and easily, and it can do lots of different jobs, like digging, dumping, hammering, raking, and even snow blowing!

Skid steer loaders use an attachment called a cold planer to repair concrete, remove bumps in pavement, and smooth surfaces so new pavement can be laid down.

The operator sits in the **CAB** in an adjustable seat.

The operator enters the cab through this **DOOR**.

Grapple forks like this one can handle all kinds of bulky, loose, and oddly shaped objects found in construction areas, demolition sites, and recycling centers.

This skid steer loader is using its angle broom to sweep the street clean.

The **BUCKET** is designed to dig, load, and carry all sorts of materials, like rocks, dirt, plants, and more.

The joystick on the right side of the cab is used to control the loader that holds the attachments. The left joystick moves the vehicle forward and backward.

The **TIRES** have extra large treads for traction on all sorts of surfaces from dirt to rock to snow to pavement.

It may be called a **MINI-EXCAVATOR**, but this small machine is mighty. It can dig, lift, carry, and dump. In fact, it can lift over 3,000 pounds (1,364 kg) and dig a hole 15 feet (4.6 meters) deep. These machines are useful for digging in areas where a larger excavator would be just too big to fit.

The hydraulic hammer works like a jackhammer to smash rocks into tiny pieces.

The end of the arm is called the **STICK**, and it holds the bucket or other attachments, like hammers or augers.

TRACKS

This mini-excavator is using its auger (a large drill) to dig a hole so a shrub can be planted.

This wide bucket is used to clear out ditches.

The **CAB** contains all the levers, pedals, and joysticks the operator needs to control the excavator.

This **BLADE** pushes dirt.

The **BUCKET** digs into the ground and scoops dirt and rocks.

The mini-excavator can lift over 3,000 pounds (1,364 kg). That's more than the weight of three cows!

The **TELEHANDLER** is used to lift things high into the air. It has a long arm called a telescopic boom that slides out like a telescope. This arm can have many different attachments at the end—like a bucket or a forklift, to name just two.

Farmers use telehandlers to transport bales of hay. Since telehandlers can carry up to 5,000 pounds (2,268 kg), this 1,500-pound (680-kg) hay bale is an easy job!

BOOM

This telehandler can stretch out its boom to lift anything—even people—as high as 55 feet (16.8 meters) in the air. That's as tall as three giraffes!

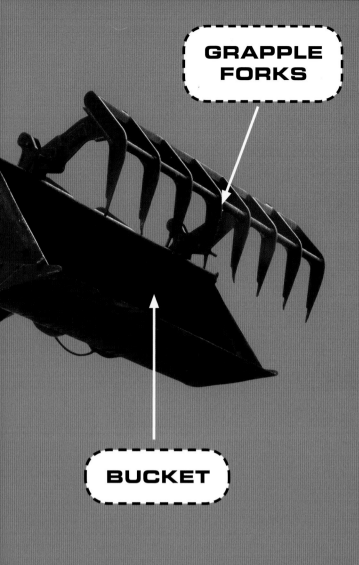

GRAPPLE FORKS

BUCKET

This telehandler is using its block fork tines (large metal prongs like those on a dinner fork) to lift a stack of concrete slabs to the very top of a new building.

The **CAB** is soundproof to protect the operator's ears.

Telehandlers use a special attachment called a grapple bucket to move and lift bulky objects like this pile of wood.

BACKHOE LOADER

The **BACKHOE LOADER** does twice the work of most machines. That's because on one end it has a loader that holds buckets, forks, and brooms. On the other end, it has a backhoe that holds buckets or hammers. This machine is a specialist at road building because it can dig, lift, and carry dirt and rocks to clear the way for new streets.

The huge teeth on this bucket help to break up the dirt as the machine digs into the ground.

CAB

LOADER BUCKET

The bucket can carry up to 9,500 pounds (4,309 kg). That's the weight of four compact cars.

The seat swivels so that the operator can face either direction in order to operate the loader on one end or the backhoe on the other.

The **BACKHOE BOOM** is the curved arm that can be used for digging, lifting, and loading.

BACKHOE BUCKET

This mechanical foot is called a **STABILIZER**, and it helps hold the backhoe loader in place while it's working.

This backhoe loader is helping clear away dirt and rocks so a new road can be built.

Have you ever ridden on a smooth, new road? Then you should thank a **MOTOR GRADER**. That's because this machine is used to build roads, parking lots, and playing fields. The motor grader has a blade that pushes dirt to make the ground flat for a new road. But that's not all a motor grader can do—it can also move snow and clear away debris on construction jobs.

CAB

The **RIPPER** can be lowered to tear through hard-packed dirt.

This motor grader is cutting grass and moving dirt to make a new road wider.

The **BLADE** scrapes the ground, pushing dirt aside.

A snow wing is a special type of extra-large blade that motor graders use to clear away snow on wintry roads.

If you were the operator of a motor grader, this is what you would see inside the cab. The levers are used to control the blade and other parts of the motor grader. The monitors above the steering wheel tell the operator when the machine is ready to go.

These **WHEELS** tilt so the motor grader can work on hillsides.

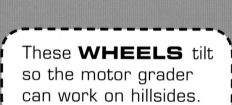

The **CIRCLE** is the piece that holds and moves the blade.

This motor grader's blade is pushing aside a pile of dirt to build a new road.

This big dump truck, called an **ARTICULATED TRUCK**, is one of the world's best earth-moving machines. It's specially built to turn in tight spaces for easy loading. This truck can carry 31 tons (28 tonnes). That means it could haul around six elephants (if they could somehow fit inside!). Some articulated trucks, like this one, can eject dirt out the back, making it easier to quickly unload the truck.

EJECTOR BLADE

The **DUMPER BODY** is the large cargo area on the back of the truck that carries the load.

EJECTOR

The excavator behind this truck is filling the truck's body with a load of dirt.

Articulated trucks use a special link called a **HITCH** that helps the truck bend in the middle so it can make sharp turns.

The ejector blade is made from high-strength steel and is used to push the load out of the dumper body.

The operator climbs up the **STAIRS** to get inside the truck's cab.

This truck is using its ejector blade to dump a load of dirt.

The truck has two seats inside its cab—one for the operator and one for a trainee who is learning how to operate the truck.

SOIL COMPACTOR

Since the **SOIL COMPACTOR** weighs 72,164 pounds (32,733 kg), it's perfect for its job—packing dirt and smoothing it to build new roads. When extra dirt is brought in to build roads higher, the soil compactor presses the dirt firmly in place to guarantee a safe, smooth road.

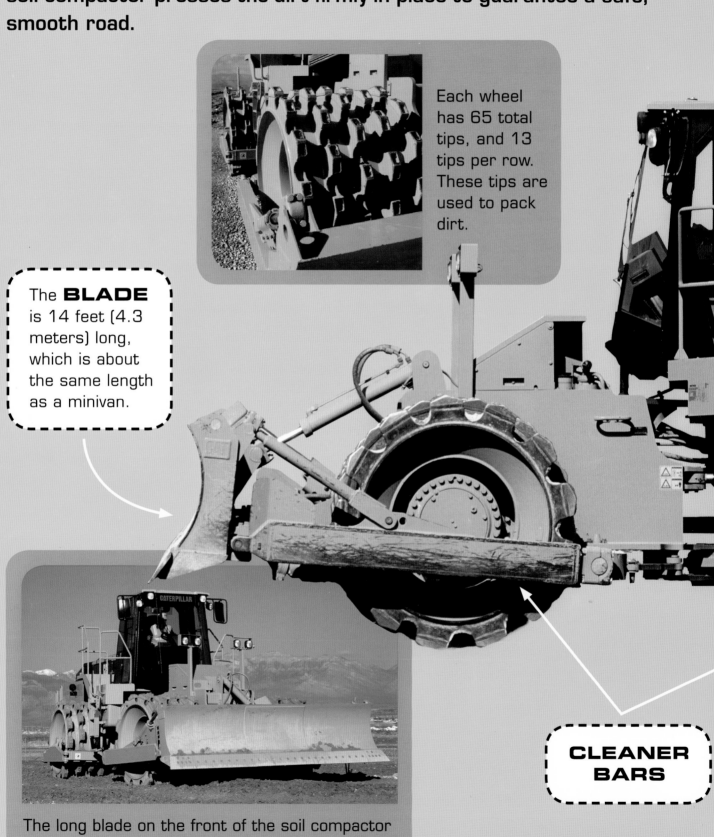

Each wheel has 65 total tips, and 13 tips per row. These tips are used to pack dirt.

The **BLADE** is 14 feet (4.3 meters) long, which is about the same length as a minivan.

CLEANER BARS

The long blade on the front of the soil compactor pushes, scrapes, and spreads dirt.

The soil compactor's **CAB** has space inside for a lunch cooler and a drink, and it also has a hook for the operator's coat.

Each of the machine's wheels has two cleaner bars that keep dirt from building up on the tips and the wheels.

The **WHEELS** are made of tough, heavy steel.

This soil compactor is busily scraping away dirt and smoothing out the ground to get it ready for a new road.

The **MATERIAL HANDLER** is great at picking up all sorts of items like metals, wires, and wood. It is often used in dumps, scrapyards, and recycling centers. The machine's long arm (actually a boom and stick) has a 50-foot (15.2-meter) reach, which means it can stretch across a basketball court!

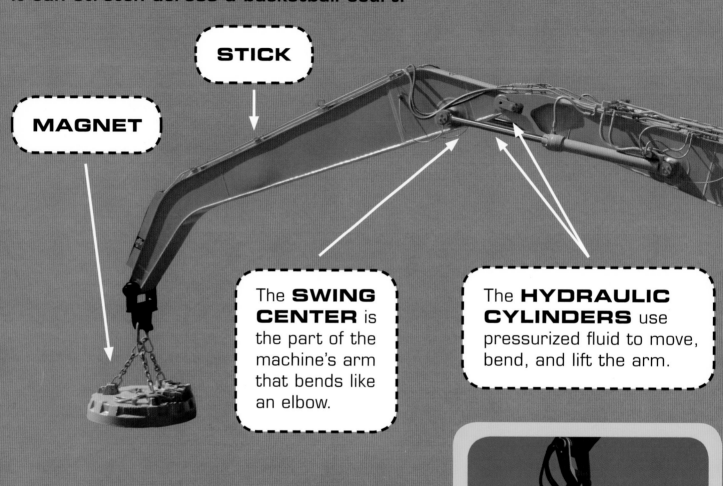

STICK

MAGNET

The **SWING CENTER** is the part of the machine's arm that bends like an elbow.

The **HYDRAULIC CYLINDERS** use pressurized fluid to move, bend, and lift the arm.

A magnet is attached to the arm of this material handler. This mighty magnet is used to pick up scrap metals.

This attachment is called an orange peel grapple because its four tines look like sections of orange peel. The strong tines can dig deeply into scrap piles. Then they grab materials and hold on tight!

The **BOOM** and the stick are the two parts of the machine's arm.

This material handler is using its orange peel grapple to lift a bundle of old wires.

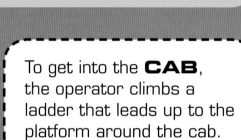

To get into the **CAB**, the operator climbs a ladder that leads up to the platform around the cab.

TRACKS

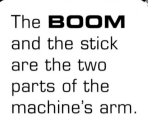

The bottom of the material handler is called the undercarriage, and it is very high off the ground so the machine can easily roll over obstacles.

BULLDOZER

The **BULLDOZER** is built for tough working conditions. It moves on crawler tracks to push mounds of rocks and dirt at construction sites. These tractors are also used in coal mines to move the *overburden* (dirt and rocks that are mixed in with the coal). They help clean up after earthquakes and tornadoes, too.

The **BLADE** bulldozes dirt, rock, and other materials. This blade is more than 9 feet (2.8 meters) tall!

This attachment, found on the back of a tractor, is called a multi-shank ripper, and its powerful teeth work like giant claws to tear up concrete or pry out rocks buried in the ground.

These four bulldozers are working hard to move mountains of dirt.

LIGHTS

This is the tractor's **ENGINE**.

Inside the tractor, the operator uses pedals, switches, and joysticks to operate the machine. A computerized control panel lets the operator know if everything is working correctly.

The crawler tracks move around this circular **SPROCKET**.

The **CRAWLER TRACKS** are giant belts that move the tractor along the ground.

Bulldozers use a special type of blade that slants down to work on hills and in ditches.

HYDRAULIC EXCAVATOR

If you thought the mini-excavator was mighty, wait until you see its big brother, the **HYDRAULIC EXCAVATOR**. This gigantic machine can dig a hole 31 feet (9.4 meters) deep and can lift over 60,000 pounds (27,216 kg). The excavator itself weighs 192,680 pounds (87,398 kg), which is as much as 48 rhinoceroses!

After digging a trench with its bucket, an excavator can also position pipes that carry water, gas, or electrical cables.

BUCKET

STICK

This attachment is called a multi-processor. It's a demolition tool whose strong jaws are used to crush and cut steel and other metals.

The **CRAWLER TRACKS** are huge belts that move the machine forward (like the way a tank moves).

The **HYDRAULIC CYLINDER** uses fluid under pressure to move and lift the bucket.

The excavator is emptying its overflowing bucket into a dump truck so the dirt can be hauled away and used elsewhere.

The **BOOM** is the long arm that holds the hydraulic cylinder.

The **ENGINE** creates the power used to run the excavator.

This excavator's hydraulic hammer is all ready to break up a slab of concrete.

The **WHEEL LOADER** is a powerful machine that weighs 430,858 pounds (195,433 kg) and is 64 feet (19.5 meters) long. That means it weighs as much as 130 cows and is as long as seven of those cows! Its sturdy lift arms and huge buckets are forceful tools for digging and lifting.

This wheel loader is getting ready to dump its full bucket of dirt.

The **LIFT ARMS** can extend 47 feet (14.3 meters) into the air. That's about the height of a four-story building!

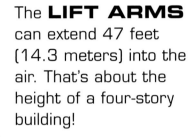

The jagged edges on the **BUCKET** work like teeth to bite into the ground.

The bucket can hold up to 35 tons (32 tonnes)—that's the weight of more than three school buses, so this load of rocks is no problem!

The load of dirt and rock spills out of the bucket and into the body of a dump truck.

The **WHEELS** are each 12 feet (3.7 meters) tall. That's higher than a professional basketball net!

The wheel loader's cab is soundproof to protect the operator's ears from the loud noises of construction.

Weighing in at 1,375,000 pounds (623,690 kg), the massive **MINING TRUCK** is one of the world's biggest hauling machines. In fact, this off-highway truck is so huge that it weighs more than all eleven of the other trucks combined and is as tall as a three-story building! This truck is so heavy that it can't drive on regular highways (it would damage them). It has to be taken in pieces on a train to the mine site and put together there!

CANOPY

The driver has to climb this set of **STAIRS** to get inside the cab.

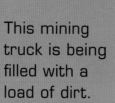

This mining truck is being filled with a load of dirt.

The **BODY** is the part of the truck that holds and hauls materials.

The mining truck can carry up to 760,000 pounds (344,730 kg), so it could carry the weight of three blue whales—the heaviest animals on Earth!

When the mining truck's body is raised up to dump its load, the body reaches 50 feet (15.2 meters) into the air.

This team of trucks is busy hauling loads of dirt.

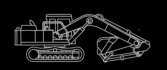   

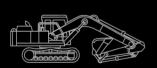